AF801047

Axolotl ADVENTURES

CURIOUS CREATURES

By Didi Zayas

ISBN 978-1-7325826-4-4 (paperback edition).
Published by Antique at Heart Press, LLC.

Axolotls

are freshwater creatures that are sometimes called "Mexican walking fish," even though they're not fish at all! They’re actually salamanders, a type of amphibian. Most amphibians live in water when they’re young and then live on land, but axolotls are special – they always stay in the water and never move onto land.

AXOLOTLS "MAGICALLY" REGROW BODY PARTS

Recently, axolotls (pronounced *ak-suh-laa-tls*) have become very popular pets, perhaps due to their incredible abilities. They can "magically" regrow lost body parts, like tails and limbs. And they are forever young, because they stay in a tadpole-like form as long as they live. Let's go on an axolotl adventure as we learn more about these amazing creatures!

Water Dragons

Axolotls have some pretty unique looks! They have long, slender bodies covered in smooth skin. Their six feathery gills stick out on the sides of their heads which make them look like pocket-sized water dragons. These astonishing features allow them to breathe under water. They have small eyes and wide, smiling mouths. They come in various colors, like pink, gold, black, gray and brown. Each axolotl is unique, so they're like little amphibian masterpieces!

AXOLOTLS BREATHE THROUGH FEATHERY GILLS

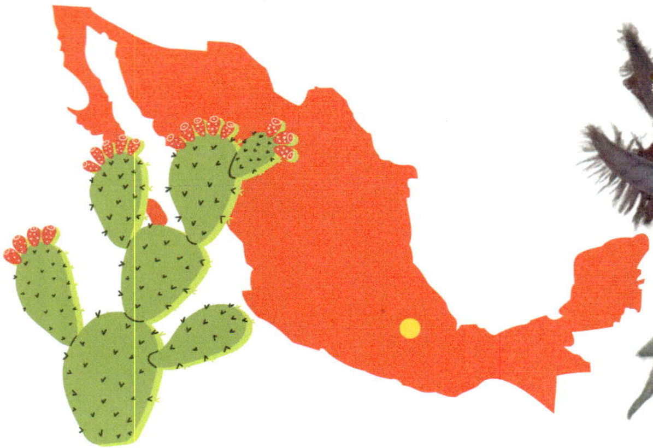

Home Sweet Home

Axolotls are only found in one place on the planet in the wild. Their habitat is a freshwater lake system in southern Mexico City (near the yellow dot on the map above). Their home, Lake Xochimilco (pronounced *sow-chee-meel-kow*), includes waterways that were dug and built by the Aztecs more than 1,000 years ago. Axolotls live on the lakebed floor, and sometimes burrow in the marshy mud to avoid predators.

Cute Carnivores

Don't let their sweet smiles fool you. Axolotls may be cute, but they're still carnivorous. That means they love to eat meat. They eat insects, earthworms and brine shrimp (seen above), and sometimes even small fish. These excellent hunters usually hunt for food at night. They use their special vacuum-like mouths to suction up prey and then swallow it whole.

THEIR MOUTHS ARE LIKE MINI VACUUMS

Stars in Their Eyes

Axolotls are nocturnal creatures, which means they're most active when the sun goes down. However, they don't really sleep like other animals do, even in the daytime. Axolotls don't have eyelids, so they can't lie down and sleep like we do. Instead, they spend most of the day resting and hiding to let their bodies refuel and relax. They have poor eyesight, so bright lights can startle these curious creatures. Their native environment is usually dark and muddy, so they're not afraid of the dark at all. They're actually more afraid of the light!

Aquatic Chameleons

Axolotls can change color due to special cells found in their skin, similar to chameleons. These skin cells contain color pigments that grow and shrink, thereby changing the hue of the salamander's skin tone. For example, an axolotl that's naturally brown or gray might turn pink or white when it feels upset. Other factors that affect their skin color include stress, age and changes to their habitat.

It’s a Wonderful Life

In the wild, axolotls usually live around 10-15 years, because they have very few predators, mainly large birds and fish. But in captivity, where they have an even safer environment, they can sometimes live up to 20 years or more! These small creatures weigh about 2-8 ounces and are about 8-12 inches in length.

LIFESPAN OF EXOTIC PETS

HEDGEHOG
5-10 YEARS

AXOLOTL
10-20 YEARS

TORTOISE
80-150 YEARS

AXOLOTLS IN INCHES

AXOLOTLS ARE USUALLY 8 INCHES LONG, SHORTER THAN A STANDARD BRUSH OR FOOTBALL

AXOLOTLS IN OUNCES

AXOLOTLS USUALLY WEIGH 2-8 OUNCES, ABOUT THE WEIGHT OF A JUICE BOX OR TENNIS BALL

Forever Young

Axolotls have a super interesting life cycle! They start out like all salamanders as tiny eggs laid in a large group. About two weeks later, they hatch into tadpoles with gills. Other salamanders go through metamorphosis and grow into their adult form. They lose their gills and begin to live on land, but not axolotls! They stay in their tadpole-like form forever. It's like they never grow up! Axolotls remain in the water, breathing through their gills for their entire lives. It's like they take a shortcut in the salamander life cycle.

Salamander Life Cycle

Legendary Animals

Long ago, the Aztecs told a story about the creation of the amazing axolotl. According to legend, all of the Aztec gods had to sacrifice themselves to create a new, better world. But Xólotl, the god of fire and death, decided to hide instead. He tried to avoid his fate by turning into a small, smooth dragon-like creature and hiding at the bottom of a lake. Sadly, the legend says the disguise failed, and Xólotl didn't survive.

But Xólotl's bravery and sacrifice made something wonderful happen. The axolotl was created in honor of him. As a result, the Aztec people thought the axolotl was really special. They used them for lots of things, like food, medicine, and special ceremonies. They felt like axolotls brought them closer to their gods.

AZTECS BELIEVED THE GODS CREATED AXOLOTLS

The Magic of Regeneration

Axolotls are like the magicians of the animal world. They have an incredible ability to regrow lost body parts in a process called regeneration. An axolotl's ability to regrow all sorts of organs and body parts has marveled scientists for generations. They can regrow their limbs and tail, as well as internal organs like their heart, lungs and even parts of their brain. They are also very resistant to many cancers. For this reason, scientists study them, hoping to find information that might help humans in the future.

Axolotls can

REGROW

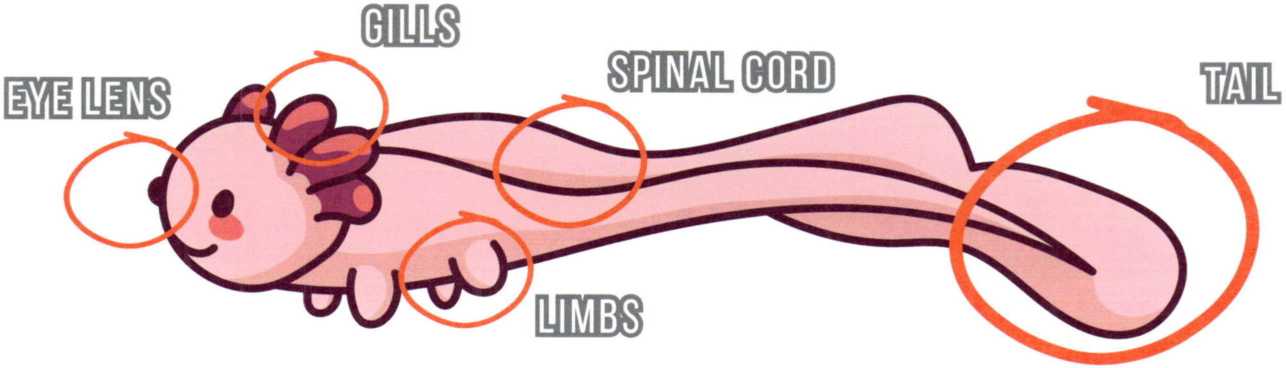

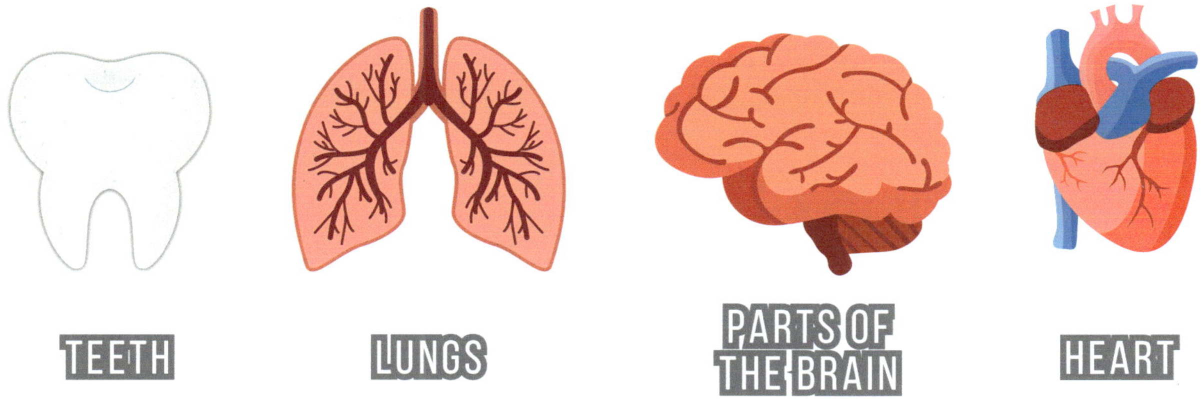

Popular Pets

Axolotls are now popular pets, but they require some special care. If you want a pet axolotl, you'll need a spacious tank with clean water, the right water temperature, and enough live or frozen prey to feed them. Taking care of an axolotl is a big responsibility. They can live a long time, so be ready for a long-term commitment!

Pet Axolotl Care Checklist

- Large Aquarium
- Tank Cooling System
- Water Filter System
- Multiple Hideouts
- Live or Frozen Prey
- Water Testing

Pop Culture Sweethearts

People are fascinated by axolotls' unique looks and abilities, and, as a result, they have risen in popularity. They're so popular that it's not uncommon to see them printed on bags, T-shirts, stickers, and now money! The government of Mexico even released a special edition bill featuring the axolotl (shown to the right). Axolotls are also well-known because they're featured as virtual pets in the video game "Minecraft."

Trouble in Paradise

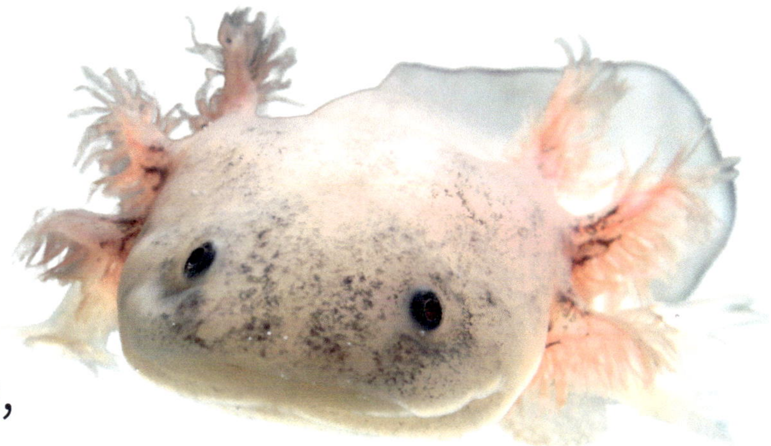

Axolotls are facing some tough times in the wild. Pollution, destruction of habitat, and new species of predatory fish have all put them at risk. Experts say that less than 1,000 axolotls exist in the wild today. At one point, scientists actually thought axolotls were extinct, but, thankfully, one was found soon after. But, don't worry about the axolotls. Scientists are working hard to protect them by studying their habits, cleaning their habitats, and breeding them in captivity to help boost their numbers.

LESS THAN 1,000 ARE LEFT IN THE WILD

EXCELLENT AXOLOTLS

Axolotls are legendary creatures, famous for their interesting looks and amazing abilities. These Mexican salamanders have been revered for thousands of years, but, in recent years, they've grown even more popular. The axolotl's powers to regrow internal organs and remain youthful forever have amazed scientists and animal lovers alike.

It's incredible how axolotls breathe through gills that make them look like dragons, change colors, and vacuum up their lunch in one big bite. It's no wonder that the world marvels at these pocket-sized "walking fish." Axolotls are endangered in the wild, but scientists are studying them closely, as they try to increase their numbers and hopefully find ways to help humans as well. There are sure to be more axolotl adventures to come as we learn more about these incredible creatures and, therefore, more about ourselves.

IMAGE CREDITS

Front Cover, above, pg. 10: ©jesue92 from Getty Images via Canva.com
Back Cover Image: ©nynkevanholten via Canva.com
Color Splashes throughout book: ©alwanuna via Canva.com
Page 1 ©williamhc from Getty Images via Canva.com
Page 2 ©kazakovmaksim from Getty Images via Canva.com
1 + 2. ©jesue92 from Getty Images via Canva.com
Page 3 ©@nynkevanholten via Canva.com
Page 4 cactus ©sketchifymexico via Canva.com
1. map ©humblino via Canva.com
2. axolotl ©Vitaliy Halenov from Getty Images via Canva.com
3. sombrero ©Vectortradition via Canva.com
4. stamp ©projectfrdmn via Canva.com
Page 5 shrimp ©S.Rohrlach from Getty Images via Canva.com
1. worm ©temmuzcan from Getty Images via Canva.com
2. axolotl ©Kaan Sezer via Canva.com
Page 6 stars ©lesliestudios via Canva.com
1. moon ©pixabay via Canva.com
2. axolotl ©Trebor Eckscher from Getty Images via Canva.com
Page 7 axolotl ©slowmotiongli from Getty Images via Canva.com
1. chameleon ©Life on White via Canva.com
Page 8 hedgehog ©nynkevanholten via Canva.com
1. axolotl ©jesue92 from Getty Images via Canva.com
2. tortoise ©Life on White via Canva.com
Page 9 football ©billionphotos via Canva.com
1. hairbrush ©chonesstock via Canva.com
2. axolotl photo ©nynkevanholten via Canva.com
3. ruler ©boldg-design via Canva.com
4. axolotl cartoon and tennis ball ©irasutoya via Canva.com
5. juice box ©sketchify via Canva.com

Page 11 tiger salamander ©Bob_Eastman via Canva.com
1. eggs ©lisad1724 from Getty Images via Canva.com
2. axolotl ©jesue92 from Getty Images via Canva.com
3. color splash ©krenstudio via Canva.com
Page 12 axolotl ©tane-mahuta from Getty Images via Canva.com
Page 13 ©sketchify via Canva.com
Page 14 axolotl ©nynkevanholten via Canva.com
1. top hat ©alexmporte0920 from pixabay via Canva.com
2. wand ©color-vectors via Canva.com (sun)
Page 15 axolotl ©sergio77 "Green Color" via Canva.com
1. tooth ©Zabi Jose from Trendify via Canva.com
2. lungs ©sparklestroke via Canva.com
3. heart ©BlenderTimer from pixabay via Canva.com
4. brain ©HtcHnm from pixabay via Canva.com
5. red circle ©thehalaldesign via Canva.com
Page 16 girl, aquarium ©Los Muertos Crew from Pexels via Canva.com
1. axolotl, grass ©Renata Tyburczy from Getty Images via Canva.com
Page 17 ©zlyka2008 from Getty Images via Canva.com
Page 18 axolotl eating snacks © ZoranMilisavljevic83 via Canva.com
1. pixelated axolotl ©Luis Madrigal from Licho via Canva.com
2. blue shirt ©WiStudio Elements via Canva.com
3. axolotl on shirt ©Olya Litvinova via Canva.com
4. Banco de Mexico 50 peso ©AmericanWildlife via istockphoto.com
Pages 19 © kazakovmaksim from Getty Images via Canva.com
Page 20 cartoon axolotl ©Selina Mai via Canva.com
1. axolotl ©Alfredo Hernandez Rios from Getty Images via Canva.com
Page 21 real axolotl ©jesue92 from Getty Images via Canva.com
1. cartoon ©ivandesign via Canva.com

Made in the USA
Monee, IL
05 November 2026

67380849R00017